J 577.586 KIN 2021

An Arctic fo

Kingsley, Imogen.

CARVER COUNTY LIBRARY

An ARCTIC FOOD WEB

BY IMOGEN KINGSLEY · ILLUSTRATED BY HOWARD GRAY

AMICUS ILLUSTRATED and **AMICUS INK**
are published by Amicus
P.O. Box 1329, Mankato, MN 56002
www.amicuspublishing.us

COPYRIGHT © 2021 Amicus. International copyright reserved in all countries. No part of this book may be reproduced in any form without written permission from the publisher.

Editor: Alissa Thielges
Designer: Kathleen Petelinsek

Library of Congress Cataloging-in-Publication Data
Names: Kingsley, Imogen, author. | Gray, Howard (Howard Willem Ian), illustrator.
Title: An arctic food web / by Imogen Kingsley ; illustrated by Howard Gray.
Description: Mankato, MN : Amicus, [2021] | Series: Amicus illustrated | Includes bibliographical references. | Audience: Ages 6-9 | Audience: Grades 2-3 | Summary: "An illustrated narrative nonfiction journey to the northern tundra that shows elementary readers how animals and plants in the Arctic ecosystem survive in an interconnected food web"— Provided by publisher.
Identifiers: LCCN 2019037112 (print) | LCCN 2019037113 (ebook) | ISBN 9781645490043 (library binding) | ISBN 9781681526461 (paperback) | ISBN 9781645490845 (pdf)
Subjects: LCSH: Tundra ecology—Arctic regions—Juvenile literature. | Food chains (Ecology)—Arctic regions—Juvenile literature.
Classification: LCC QH541.5.T8 M45 2021 (print) | LCC QH541.5.T8 (ebook) | DDC 577.5/8609113—dc23
LC record available at https://lccn.loc.gov/2019037112
LC ebook record available at https://lccn.loc.gov/2019037113

Printed in the United States of America.

HC 10 9 8 7 6 5 4 3 2 1
PB 10 9 8 7 6 5 4 3 2 1

About the Author
Imogen Kingsley has written over 20 books for children. Imogen loves animals and at one point had 3 donkeys, 3 horses, 15 chickens, 2 ducks, 9 cats, 3 pigs, 2 goats, and 1 dog. Imogen currently resides in Colorado with her family where she spends a little bit of every day watching animals in their natural habitats.

About the Illustrator
Howard Gray has illustrated a selection of fiction and non-fiction children's books. He has always considered himself an artist, but with a PhD in dolphin genetics, he has a background in zoology. He is now pursuing his dream career in children's illustration from the picturesque city of Durham, UK. Find out more at www.howardgrayillustrations.com.

Welcome to the one of the coldest ecosystems on Earth—the Arctic tundra! For much of the year, the land is covered in ice and snow. Life here makes up a unique food web. Let's see how energy moves between the plants and animals.

The food web starts in the summer. It is a very short time of year this far north. It warms just enough for the snow and ice to melt. Shrubs sprout. Grasses grow. Arctic poppies bloom. As plants grow, more animals migrate here.

5

Plants are producers. They make, or produce, their own food. To do this, they need energy from the sun and nutrients from the ground.

Luckily, the sun never sets in the summer. It is light outside 24 hours a day. The plants grow quickly with all the light.

Animals are consumers. They eat, or consume, other living things for energy. Herbivores eat only plants. They are called primary consumers.

Butterflies and moths drink nectar from flowers. Nibble. Nibble. A small lemming eats grass. So do the muskox. A huge herd of caribou munches on grass, flowers, and lichens.

9

Birds fly down and peck at the insects. The swallow and lark are secondary consumers. Bugs eat the plants and birds eat the bugs. But birds can also be someone else's meal.

A snowy owl swoops in. It is an apex predator. It sets its eyes on the lemming. But look! An Arctic fox is here, too. It also wants the lemming for lunch.

11

WHOOSH! The owl moves fast. It grabs the lemming with its talons. The fox is out of luck—for now. It will look for another meal.

Everything in the food web is connected. If the grass didn't grow one summer, the herbivores wouldn't have any food. They would leave the Arctic. Then the owl and fox would go hungry.

When winter comes again, most animals flee to warmer weather. But some stay. Lemmings live in burrows. They stay warm underground. That doesn't stop the fox from finding them, though.

In winter, plants are buried under the snow. Animals must dig to get to them. Caribou have thick hooves. They dig for lichens. They also keep an eye out for apex predators, like wolves.

15

Farther north is the cold Arctic Ocean. Here, sea animals fight for survival against land animals. The mighty polar bear hunts on the sea ice. This giant apex predator feasts on seals. Seal fat, called blubber, is the polar bear's main food.

17

18

When an animal dies, its body is picked over. Arctic animals aren't picky eaters. Food is scarce, so most animals are scavengers. Foxes and ravens follow polar bears and wolves. They eat their leftovers. Then decomposers, like bacteria, break down the rest.

20

Is the food web done? Not quite. The nutrients from the dead bodies go back into the earth. They help feed new plants in the summer. The cycle continues over and over, helping life thrive in the Arctic.

An ARCTIC FOOD WEB

apex predator An animal that has no natural predators; it is the top consumer in a food web.

secondary consumer An animal that can eat both plants and other animals, but also gets eaten by other animals.

primary consumer An animal that eats only plants and is eaten by other animals.

producer A plant that makes its own food.

scavengers and **decomposers** Animals that break down dead animals and plants by eating it, which help nutrients go back into the ground.

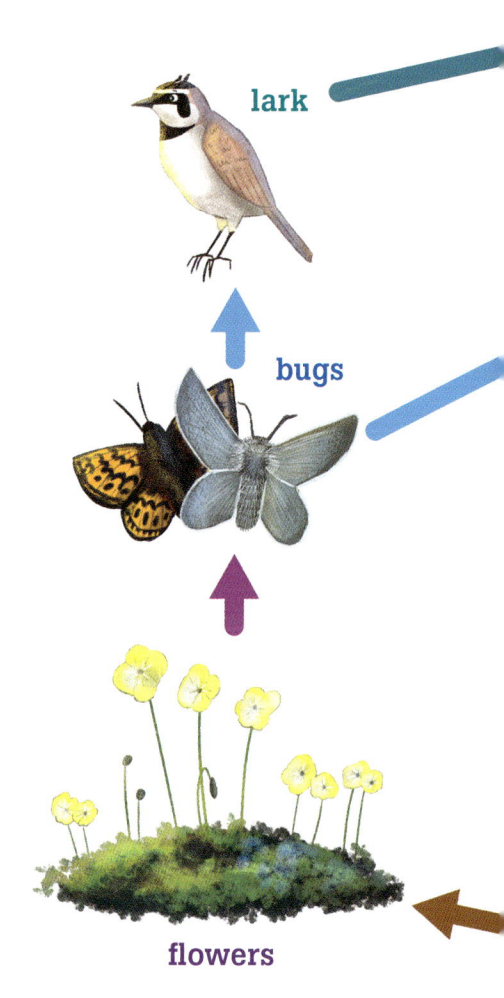

lark

bugs

flowers

snowy owl

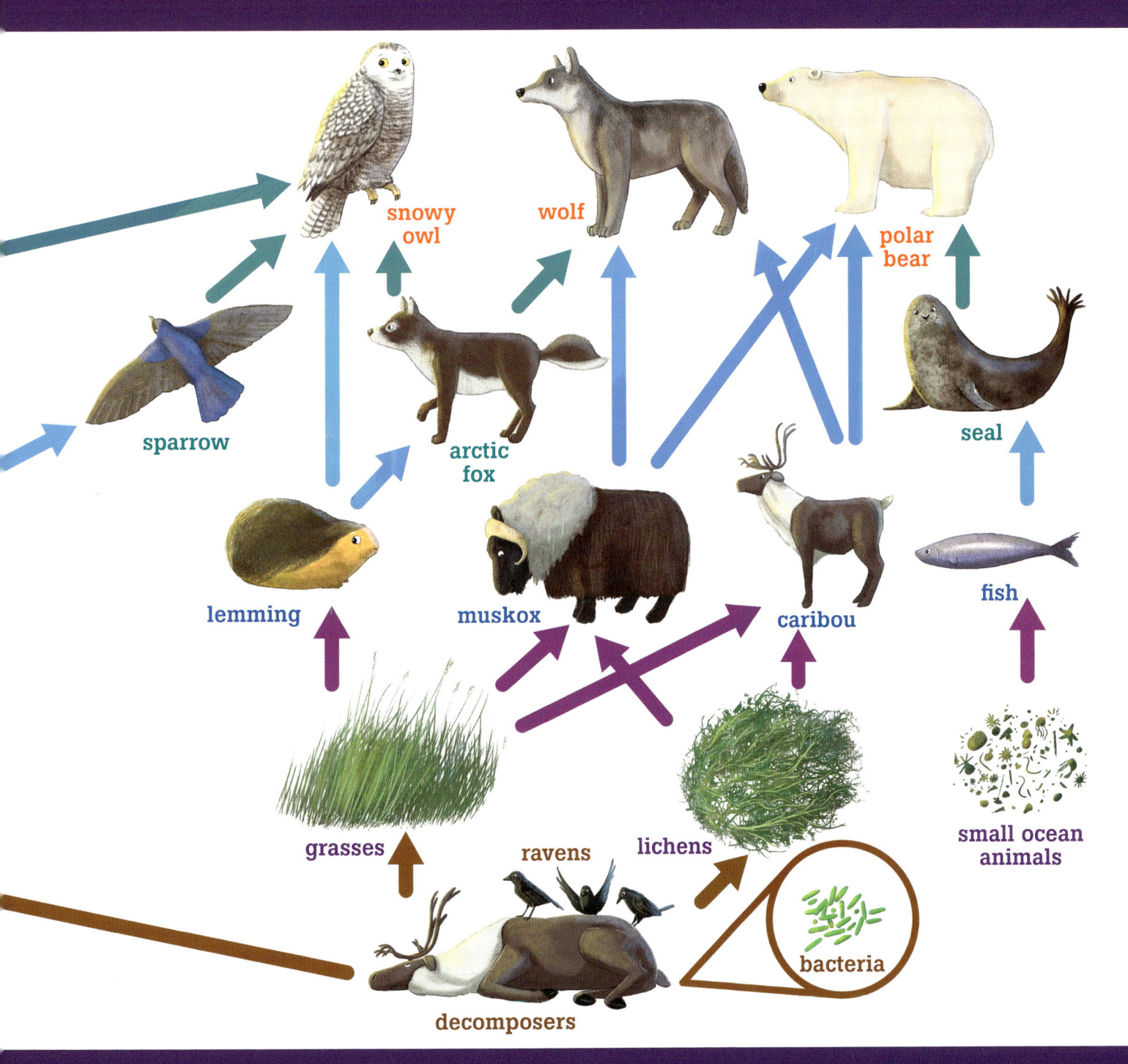

wolf

polar bear

sparrow

arctic fox

seal

lemming

muskox

caribou

fish

grasses

ravens

lichens

small ocean animals

bacteria

decomposers

23

GLOSSARY

ecosystem A community of plants and animals that live in a certain area.

food web A system of how animals and plants in an ecosystem relate; it shows who eats what.

migrate When an animal moves from one place to another, usually to the same place each year.

nutrient A substance that plants and animals need to live and grow.

tundra A cold, treeless area where part of the soil stays frozen all year.

WEBSITES

DK Find Out! The Arctic
https://www.dkfindout.com/us/animals-and-nature/habitats-and-ecosystems/arctic/

Home Sweet Habitat & Food Webs
https://thekidshouldseethis.com/post/home-sweet-habitat-food-webs-crash-course-kids

Every effort has been made to ensure that these websites are appropriate for children. However, because of the nature of the Internet, it is impossible to guarantee that these sites will remain active indefinitely or that their contents will not be altered.

READ MORE

Dunne, Abbie. **Food Chains and Webs**. North Mankato, Minn.: Capstone, 2017.

McCarthy, Cecilia Pinto. **Tundra Ecosystems**. Mankato, Minn.: 12 Story Library, 2018.

Pettiford, Rebecca. **Arctic Food Chains: Who Eats What?** Minneapolis: Jump!, 2016.